International
Association
of Fire Chiefs

National
Fire Protection
Association

Fire Officer
I, II, III, & IV
TASK BOOKLET

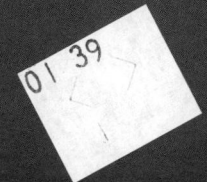
01 39

JONES & BARTLETT
LEARNING

World Headquarters
Jones & Bartlett Learning
5 Wall Street
Burlington, MA 01803
978-443-5000
info@psglearning.com
www.psglearning.com

International Association of Fire Chiefs
4025 Fair Ridge Drive
Fairfax, VA 22033
www.IAFC.org

National Fire Protection Association
1 Batterymarch Park
Quincy, MA 02169-7471
www.NFPA.org

Jones & Bartlett Learning books and products are available through most bookstores and online booksellers. To contact the Jones & Bartlett Learning Public Safety Group directly, call 800-832-0034, fax 978-443-8000, or visit our website, www.psglearning.com.

Editorial Credits
Author: Kevin Grebinar

Production Credits
VP, Product Management: Christine Emerton
Senior Managing Editor: Donna Gridley
Director - Product Management: Fire: Bill Larkin
VP, Sales: Phil Charland
Manager, Project Management: Kristen Rogers
Project Specialist: Meghan McDonagh
Digital Project Specialist: Rachel DiMaggio
Director of Marketing Operations: Brian Rooney
Production Services Manager: Colleen Lamy
VP, Manufacturing and Inventory Control: Therese Connell
Composition: S4Carlisle Publishing Services
Project Management: S4Carlisle Publishing Services
Cover Design: Scott Moden
Text Design: Scott Moden
Rights Specialist: Liz Kincaid
Cover Image: © Jones & Bartlett Learning; Title Page, Chapter Opener: © Jones & Bartlett Learning. Photographed by Glen E. Ellman.
Printing and Binding: Sheridan/CJK Group

ISBN: 978-1-284-50503-0

Printed in the United States of America
24 23 22 21 20   10 9 8 7 6 5 4 3 2 1

Jones & Bartlett Learning. Photographed by Glen E. Ellman

Contents

Preface

The task sheets and associated job performance requirements (JPRs) included in this task booklet are based on the 2020 Edition of *NFPA 1021: Standard for Fire Officer Professional Qualifications*. Together, these will allow supervising officers to track and evaluate completion of the assigned task by the candidate. The candidate's progress in gaining the experience necessary to achieve competency in future officer role(s) can be properly documented with the use of this task booklet.

It is expected that specific guidelines for the development of JPR assignments, time to completion, and any prerequisites required to enter this phase of training would be developed by the authority having jurisdiction. These items can vary based on the department's job/position expectations, local/state regulations, or department structure. When properly used, this task booklet, together with *Fire Officer: Principles and Practice, Fourth Edition* and *Chief Officer: Principles and Practice, Third Edition* curriculums, provides a comprehensive training program for future fire service officers that addresses both the education and experience necessary to be successful.

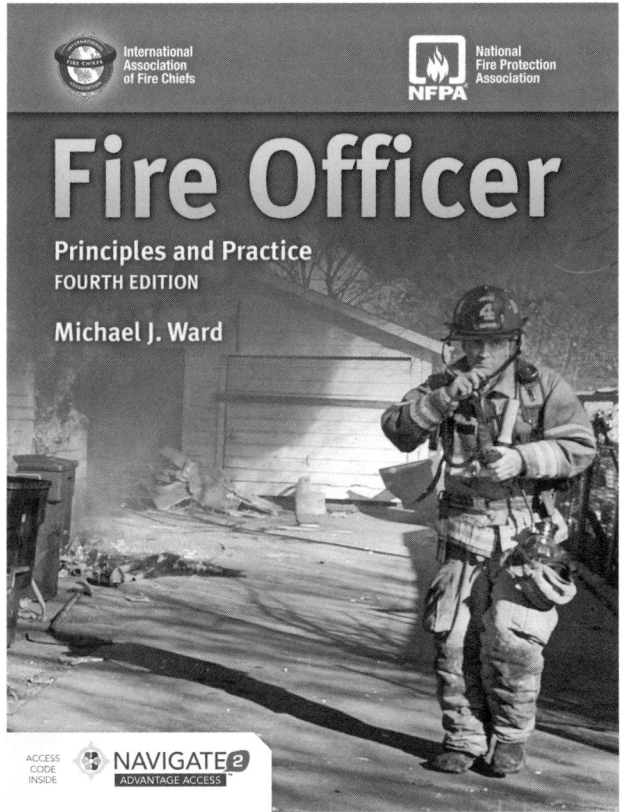

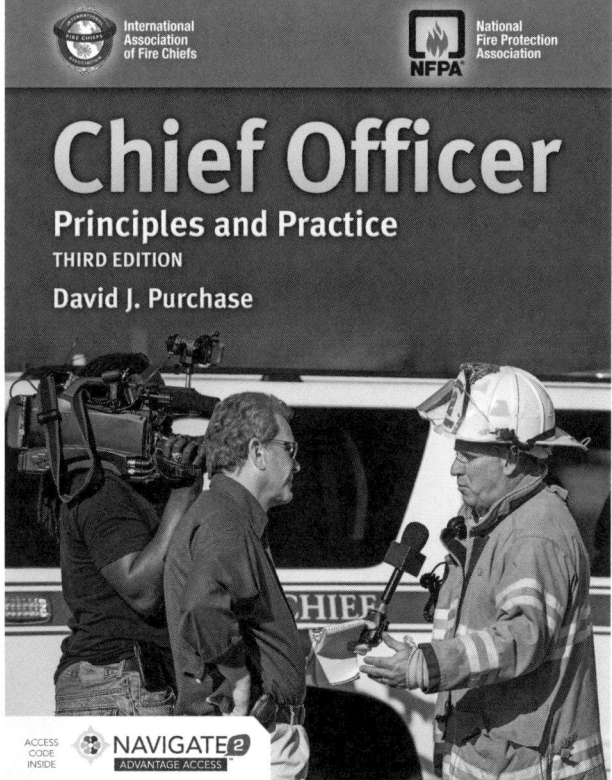

Instructions for the Fire Officer Task Booklet Evaluation Record

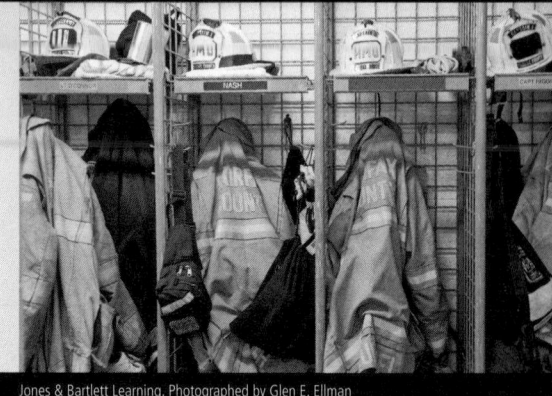

Candidate Information

Print the candidate's name, address, phone number, email address, identification number, and department affiliation.

Course and Task Booklet Completion

Record the date of course completion, the date the task booklet was initiated, completion date, and certification date.

Evaluator Information

Sign and print name to authenticate your recommendations.

Date

Document the date the Evaluation Record was completed.

Evaluator's Score

Include the date, your evaluation score, and your initials in the far-right boxes next to each numbered task to indicate that the candidate completed and passed that task. If the candidate failed the task, place an F in that box to indicate FAILED.

Record additional remarks, recommendations, or comments on an Individual Performance Evaluation or by attaching an additional sheet to the Evaluation Record.

Evaluator's Relevant Qualification (or agency certification)

List your qualification or certification relevant to the candidate's position you supervised.

Note: Evaluators must be either qualified in the position being evaluated or supervise the candidate.

Final Evaluators must be qualified in the candidate position they are evaluating.

Fire Officer I
Task Booklet

Candidate's Name _____
           First   Middle   Last

Candidate's Address _____
            Street   City   State   Zip Code

Candidate's Phone Number: _____

Candidate's Email Address: _____

Candidate's Identification #: _____

Department Affiliation: _____

Date Fire Officer I Course Completion: _____

Date Task Booklet was Initiated: _____

Date Task Booklet was Completed: _____

Certification Date: _____

Evaluator:

Signature of Evaluator _____

Printed Name of Evaluator _____

Date: _____

Evaluator's Relevant Qualification (or agency certification)

_____

_____

_____

_____

_____

_____

_____

_____

_____

1

Candidate Name: _____   Date: _____

TASK SHEET: 4.2.1, 4.4.5, 4.6.3 NFPA 1021, 2020 General Requirements	**Task:** Assign tasks or responsibilities to unit members, given an assignment at an emergency operation, so that the instructions are complete, clear, and concise; safety considerations are addressed; and the desired outcomes are conveyed.
	Task: Explain the needs and benefits of collecting incident response data, given the goals and mission of the organization, so that incident response reports are timely and accurate.
	Task: Develop and conduct a postincident analysis, given single unit incident and postincident analysis policies, procedures, and forms, so that all required critical elements are identified and communicated and the approved forms are completed and processed in accordance with policies and procedures.
PERFORMANCE OUTCOME:	The candidate will assume the role of company officer supervising the first-due fire company at a residence fire (actual or simulated). Candidate will assign tasks or responsibilities in a complete, clear, and concise manner so that safety considerations are addressed and desired outcomes are conveyed. The candidate will collect all incident response data and complete all organizational incident response forms. The candidate will then conduct a postincident analysis of the incident, using proper policies and procedures.

EQUIPMENT REQUIRED: Firefighting equipment necessary to complete evolutions.

CONDITIONS: The candidate will complete all elements of the assigned task. Include candidate's narrative on task completion, forms, photos/drawings, etc., department policy or procedure.

No.	Task Steps	Date	Evaluator Score	Evaluator Initials
1.	Assign tasks or responsibilities to unit members at an emergency.			
2.	Condense instructions in an understandable way.			
3.	Give instructions that are complete, clear, and concise.			
4.	Confirm understanding of assignments.			
5.	Convey desired outcomes.			
6.	Efficiently use personnel and equipment available to the company.			
7.	Conduct an incident response report using proper policies, forms, and procedures.			
8.	Conduct a postincident analysis using proper policies, forms, and procedures.			

Evaluator: _____   _____
                            Printed   **Signed**

Comments:

_____

_____

Candidate Name: _____ Date: _____

TASK SHEET: 4.2.2, 4.2.6 NFPA 1021, 2020 General Requirements	**Task:** Assign tasks or responsibilities to unit members, given an assignment under nonemergency conditions at a station or other work location, so that the instructions are complete, clear, and concise; safety considerations are addressed; and the desired outcomes are conveyed.
	Task: Coordinate the completion of assigned tasks and projects by members, given a list of projects and tasks and the job requirements of subordinates, so that the assignments are prioritized, a plan for the completion of each assignment is developed, and members are assigned to specific tasks and supervised during the completion of the assignments.
PERFORMANCE OUTCOME:	The candidate will assume the role of company officer supervising other fire fighters at a station. The candidate will assign nonemergent job duties or projects to unit members, such as station duties, apparatus maintenance, and special projects. The assignment will be to assign specific tasks and resources to each individual fire fighter, while providing for adequate supervision and safety considerations, so that the company's assignment is completed. Make a written plan of what specific tasks and resources are assigned to each fire fighter. Establish an order of priority of tasks and a timeline for completion. The company will remain run-ready at all times.

EQUIPMENT REQUIRED: Paper, pen/pencil, computer, if applicable.

CONDITIONS: The candidate will complete all elements of the assigned task. Include candidate's narrative on task completion, forms, photos/drawings, etc., department policy or procedure.

No.	Task Steps	Date	Evaluator Score	Evaluator Initials
1.	Establish a reliable method of alerting company for nonemergency assignments.			
2.	Provide appropriate safety equipment to each member based on task.			
3.	Give instructions that are clear, concise, and precise.			
4.	Efficiently use personnel and equipment available to the company.			
5.	Provide for adequate supervision of each member.			
6.	Create a written plan that fully accomplishes the assignment.			

Evaluator: _____   _____
　　　　　　　　　　　Printed　　　　　　　　　　　　　　　　　　　　　　　　　　**Signed**

Comments:

_____

_____

Candidate Name: _____ Date: _____

TASK SHEET: 4.2.3 NFPA 1021, 2020 **General Requirements**	**Task:** Direct unit members during a training evolution, given a company training evolution and training policies and procedures, so that the evolution is performed safely, efficiently, and as directed.			
PERFORMANCE OUTCOME:	The candidate will assume the role of a company officer conducting a training evolution for those under his or her command. The candidate will communicate verbal instructions to the company so that the evolution is safely and efficiently performed according to applicable policy and procedures.			
EQUIPMENT REQUIRED: Company members, training equipment necessary to complete the assigned evolution. Policies and procedures.				
CONDITIONS: The candidate will complete all elements of the assigned task. Include candidate's narrative on task completion, forms, photos/drawings, etc., department policy or procedure.				

No.	Task Steps	Date	Evaluator Score	Evaluator Initials
1.	Provide written or verbal instructions to the company members.			
2.	Conduct the evolution in a safe and efficient manner.			
3.	Ensure compliance with applicable policies and procedures.			
4.	Maximize learning by anticipating needs or problems.			
5.	Address improper methods or mistakes made by company members.			
6.	Successfully complete the training evolution.			

Evaluator: _____ _____
                                  Printed   **Signed**

Comments:

_____

_____

Candidate Name: _____   Date: _____

TASK SHEET: 4.2.4, 4.2.5 NFPA 1021, 2020 General Requirements	**Task:** Recommend action for member-related problems, given a member with a situation requiring assistance and the member assistance policies and procedures, so that the situation is identified and the actions taken are within the established policies and procedures.
	Task: Apply human resource policies and procedures, given an administrative situation requiring action, so that policies and procedures are followed.

PERFORMANCE OUTCOME:	The candidate will assume the role of company officer. A subordinate member of the fire department approaches the company officer with a problem. Member-related problems could include substance abuse; acute, chronic, and delayed stress; and health, financial, personal, family, and other situations that may adversely affect the member's job performance. Candidate must listen carefully to determine the true nature of the problem and provide emotional support through active listening. Candidate will determine an initial course of action (within the company officer's scope of authority), explain the course of action to the member, and make appropriate verbal notification and written documentation to the company officer's next-in-line supervisor.

EQUIPMENT REQUIRED: Subordinate member and supervisor of the company officer. Policies and procedures. Paper, pen/pencil, computer, if applicable.

CONDITIONS: The candidate will complete all elements of the assigned task. Include candidate's narrative on task completion, forms, photos/drawings, etc., department policy or procedure.

No.	Task Steps	Date	Evaluator Score	Evaluator Initials
1.	Ensure the privacy of conversation between officer and subordinate.			
2.	Understand and apply knowledge of post-critical incident stress, and/or other stress-related situations.			
3.	Demonstrate a caring, mature, and responsible attitude.			
4.	Adhere to applicable policies and procedures.			
5.	Provide written notification to officer's supervisor as soon as possible.			
6.	Provide a good faith written recommendation for further action to officer's supervisor.			

Evaluator: _____   _____
Printed   **Signed**

Comments:

_____

_____

Candidate Name: _____   Date: _____

TASK SHEET: 4.3.1 **NFPA 1021, 2020** **General Requirements**	**Task:** Initiate action on a community need, given policies and procedures, so that the need is addressed.

PERFORMANCE OUTCOME:	Candidate will appropriately respond to a routine request from a citizen of the community (e.g., safety talk, safety drill, car seat inspections, neighborhood request, etc.). Candidate will answer the need accurately, courteously, and in accordance with policies and procedures. Candidate will then initiate the process and respond to the desired community need.

EQUIPMENT REQUIRED:

CONDITIONS: The candidate will complete all elements of the assigned task. Include candidate's narrative on task completion, forms, photos/drawings, etc., department policy or procedure.

No.	Task Steps	Date	Evaluator Score	Evaluator Initials
1.	Demonstrate understanding/compliance with policies and procedures.			
2.	Respond to the community need accurately and in a timely fashion.			
3.	Demonstrate the ability to coordinate and schedule a community need.			
4.	Provide or deliver resources to meet the requested community need.			

Evaluator: _____   _____
                              Printed   **Signed**

Comments:

_____

_____

Candidate Name: _____   Date: _____

TASK SHEET: 4.3.2 **NFPA 1021, 2020** **General Requirements**	**Task:** Initiate action to a citizen's concern, given policies and procedures, so that the concern is answered or referred to the correct individual for action and all policies and procedures are met with compliance.
PERFORMANCE OUTCOME:	The candidate will assume the role of a company officer when a citizen comes to the fire station to make a complaint (e.g., careless driving, excessive noise of fire department vehicles or training in the neighborhood, etc.). Candidate will receive the complaint, provide an immediate verbal response to satisfy the citizen's desire that something is to be done, and follow up by initiating proper action according to policy.

EQUIPMENT REQUIRED: Citizen and policies and procedures.

CONDITIONS: The candidate will complete all elements of the assigned task. Include candidate's narrative on task completion, forms, photos/drawings, etc., department policy or procedure.

No.	Task Steps	Date	Evaluator Score	Evaluator Initials
1.	Give serious and immediate attention to the citizen's complaint.			
2.	Assure citizen that action will be taken to alleviate the concern.			
3.	Behave in a respectful, professional, and courteous manner.			
4.	Allow the citizen time to adequately communicate the concern.			
5.	Initiate proper action as required by policy.			
6.	Make notification of complaint to the proper individual, if applicable.			

Evaluator: _____   _____
                                    Printed   **Signed**

Comments:

_____

_____

Candidate Name: _____   Date: _____

TASK SHEET: 4.3.3 NFPA 1021, 2020 General Requirements	**Task:** Respond to a public inquiry, given the policies and procedures, so that the inquiry is answered accurately, courteously, and in accordance with applicable policies and procedures.
PERFORMANCE OUTCOME:	Candidate will assume the role of a fire department officer and respond to a public inquiry (e.g., application processes, inspection/code questions, permits, etc.). Candidate will answer the inquiry accurately, courteously, and according to established policies and procedures.

EQUIPMENT REQUIRED: Policies or procedures. Department forms, pen, paper, computer.

CONDITIONS: The candidate will complete all elements of the assigned task. Include candidate's narrative on task completion, forms, photos/drawings, etc., department policy or procedure.

No.	Task Steps	Date	Evaluator Score	Evaluator Initials
1.	Demonstrate understanding/compliance with policies and procedures.			
2.	Answer the public inquiry accurately.			
3.	Project a professional and courteous demeanor.			
4.	Demonstrate ability to effectively communicate verbally.			
5.	Demonstrate effective written communication, if applicable.			
6.	Respond to the public inquiry in a timely fashion.			

Evaluator: _____   _____
                              Printed   **Signed**

Comments:

_____

_____

Candidate Name: _____   Date: _____

TASK SHEET: 4.4.1, 4.4.2 NFPA 1021, 2020 **General Requirements**	**Task:** Recommend changes to existing departmental policies and/or implement a new departmental policy at the unit level, given a new departmental policy, so that the policy is communicated to and understood by unit members. **Task:** Execute routine unit-level administrative functions, given forms and record management systems, so that the reports and logs are complete and files are maintained in accordance with policies and procedures.
PERFORMANCE OUTCOME:	Candidate will assume the role of a company officer and recommend change to existing policy or implement new policy that needs to be established. Following the approval of such policy, the candidate will provide this policy as new information and provide an implementation to the company training members. Candidate will then update or make changes to any report forms, logs, or filing systems that are affected by the implementation of the new or revised policy.

EQUIPMENT REQUIRED: Company members. Policy covering written reports of any type. Forms or reports required by the policy. Pen/pencil, computer, if applicable.

CONDITIONS: The candidate will complete all elements of the assigned task. Include candidate's narrative on task completion, forms, photos/drawings, etc., department policy or procedure.

No.	Task Steps	Date	Evaluator Score	Evaluator Initials
1.	Show understanding and personal compliance with new policy.			
2.	Describe new policy in a manner understandable to the members.			
3.	Answer questions correctly with regard to the new policy, if any.			
4.	Demonstrate how new policy requires form/reports to be completed.			
5.	Adequately communicate information verbally and in writing.			
6.	Communicate why the new policy is necessary.			

Evaluator: _____   _____
          Printed   **Signed**

Comments:

_____

_____

Candidate Name: _____   Date: _____

TASK SHEET: 4.4.3	**Task:** Prepare a budget request, given a need and budget forms, so that the request is in the proper format and is supported with data.
NFPA 1021, 2020	
General Requirements	

PERFORMANCE OUTCOME:	Candidate will create a budget request using the proper forms, procedures, and supporting data, and then submit the budget to the proper budget coordinator.

EQUIPMENT REQUIRED: Specific facility. Pen/pencil and paper. Necessary budget forms or reports. Budget planning policy and procedures. Uniform or other credentials to ensure proper identification to business owners/occupants when obtaining supporting budget data. Computer, if applicable.

CONDITIONS: The candidate will complete all elements of the assigned task. Include candidate's narrative on task completion, forms, photos/drawings, etc., department policy or procedure.

No.	Task Steps	Date	Evaluator Score	Evaluator Initials
1.	Obtain proper request forms and procedures.			
2.	Research revenue sources for budget.			
3.	Obtain supporting data to the budget request.			
4.	Develop and organize an outlined budget plan.			
5.	Produce completed plan using the appropriate forms and reports.			
6.	Submit complete budget packet to proper budget coordinator.			

Evaluator: _____   _____
                                Printed   **Signed**

Comments:

_____

_____

Candidate Name: _____   Date: _____

TASK SHEET: 4.4.4 NFPA 1021, 2020 General Requirements	**Task:** Explain the purpose of each management component of the organization, given an organization chart, so that the explanation is current and accurate and clearly identifies the purpose and mission of the organization.
PERFORMANCE OUTCOME:	Candidate will provide a current copy of their department's organizational chart with defined responsibilities and duties, then make recommended changes to that organizational chart that would improve the efficiency of their organization. All changes must have written justification. If no changes are identified, written reinforcement to the organizational structure must be created. If the candidate has no organizational chart in their department, they will create one with written defined responsibilities and duties.

EQUIPMENT REQUIRED:

CONDITIONS: The candidate will complete all elements of the assigned task. Include candidate's narrative on task completion, forms, charts and defined responsibilities etc., department policy or procedure.

No.	Task Steps	Date	Evaluator Score	Evaluator Initials
1.	Identify the structure of an organization.			
2.	Identify the functions of management.			
3.	Communicate in writing the mission of the organization.			
4.	Communicate in writing the defined responsibilities and duties of the organization.			
5.	Correct, reinforce, or develop defined management components of an organization.			

Evaluator: _____   _____
                                Printed   **Signed**

Comments:

_____

_____

Candidate Name: _____   Date: _____

| TASK SHEET: 4.5.1, 4.5.2

NFPA 1021, 2020

General Requirements | **Task:** Describe the procedures of the AHJ for conducting fire inspections, given any of the following occupancies, so that all hazards, including hazardous materials, are identified, approved forms are completed, and approved actions are taken.

1. Assembly
2. Educational
3. Health care
4. Detention and correctional
5. Residential
6. Mercantile
7. Business
8. Industrial
9. Storage
10. Unusual structures
11. Mixed occupancies

Task: Identify construction, alarm, detection, and suppression features that contribute to or prevent the spread of fire, heat, and smoke throughout the building or from one building to another, given an occupancy and the policies and forms of the AHJ so that a preincident plan for any of the following occupancies is developed.

1. Public assembly
2. Educational
3. Institutional
4. Residential
5. Business
6. Industrial
7. Manufacturing
8. Storage
9. Mercantile
10. Special properties |
| **PERFORMANCE OUTCOME:** | Candidate will assume the role of a company officer and will conduct a fire inspection of one of the occupancies listed above. All findings of the inspection shall be documented in accordance with approved policies and procedures of the AHJ. In addition, the candidate will identify construction, alarm, detection, and suppression systems and develop a pre-incident plan for the occupancy in accordance with approved policies and procedures of the AHJ. |

EQUIPMENT REQUIRED: Specific facility. Transportation to/from assigned facility. Pen/pencil and paper. Necessary inspection and pre-incident plan forms or reports. Inspection and pre-incident planning policy and procedures. Uniform or other credentials to ensure proper identification to business owners/occupants when obtaining inspection and pre-incident plan data. Computer, if applicable.

CONDITIONS: The candidate will complete all elements of the assigned task. Include candidate's narrative on task completion, forms, photos/drawings, etc., department policy or procedure.

No.	Task Steps	Date	Evaluator Score	Evaluator Initials
1.	Initiate initial contract with courtesy and professionalism.			
2.	Obtain cooperation by emphasizing the reasoning behind the inspection and pre-incident plan.			
3.	Exhibit professional appearance and demeanor for the site visit.			
4.	Include all elements of the fire inspection according to policy. Forms to include site-specific hazards and hazardous materials.			
5.	Include all elements of the pre-incident plan according to policy, forms, drawings, etc.			
6.	Produce a completed fire inspection document using the appropriate forms and reports.			
7.	Produce completed plan using the appropriate forms and reports.			
8.	Communicate effectively using both verbal and written methods.			

Evaluator: _____   _____
                                  Printed   **Signed**

Comments:

_____

_____

Candidate Name: _____   Date: _____

TASK SHEET: 4.5.3 NFPA 1021, 2020 **General Requirements**	**Task:** Secure an incident scene, given rope or barrier tape, so that unauthorized persons can recognize the perimeters of the scene, are kept from restricted areas, and all evidence or potential evidence is protected from damage or destruction.
PERFORMANCE OUTCOME:	Candidate will assume the role of fire department officer and will be given a real or simulated fire incident scene. The candidate will identify a preliminary need for a fire investigation and secure the scene and evidence by establishing perimeters to the scene. Candidate will identify potential witnesses and demonstrate the proper procedure for calling an investigator.

EQUIPMENT REQUIRED: Real or simulated fire incident scene with materials necessary to create the proper environment. Rope or barrier tape. Applicable reports or witness statement forms. Persons to act as first-arriving members and others such as witnesses, occupants, or others with incident information.

CONDITIONS: The candidate will complete all elements of the assigned task. Include candidate's narrative on task completion, forms, photos/drawings, etc., department policy or procedure.

No.	Task Steps	Date	Evaluator Score	Evaluator Initials
1.	Identify the need for a fire investigation.			
2.	Adequately secure the fire scene to protect evidence.			
3.	Establish a scene perimeter that prohibits unauthorized entry.			
4.	Identify potential witnesses.			
5.	Establish need for investigator and use proper methods to request one.			

Evaluator: _____   _____
                              Printed   **Signed**

Comments:

_____

_____

Candidate Name: _____   Date: _____

TASK SHEET: 4.6.1, 4.6.2, 4.6.3 NFPA 1021, 2020 **General Requirements**	**Task:** Develop an initial action plan, given size-up information for an incident and as-signed emergency response resources, so that resources are deployed to control the emergency. **Task:** Implement an action plan at an emergency operation, given assigned resources, type of incident, and a preliminary plan, so that resources are deployed to mitigate the situation. **Task:** Develop and conduct a postincident analysis, given a single unit incident and postincident analysis policies, procedures, and forms, so that all required critical elements are identified and communicated and the approved forms are completed and processed in accordance with policies and procedures.
PERFORMANCE OUTCOME:	Candidate will develop and implement an initial action plan for an emergency incident scenario. Candidate must be able to analyze emergency scene conditions, allocate resources, communicate verbally and in writing, operate within an emergency management system, and supervise and account for assigned personnel so that resources are effectively deployed to mitigate the situation. The candidate will then conduct a postincident analysis using proper policies and procedures.

EQUIPMENT REQUIRED: Emergency incident scenario including type of incident, size-up information, and assigned resources. Policies and procedures. Pen/pencil and paper. Necessary forms and reports. Personnel accountability system components. Computer, if applicable.

CONDITIONS: The candidate will complete all elements of the assigned task. Include candidate's narrative on task completion, forms, photos/drawings, etc., department policy or procedure.

No.	Task Steps	Date	Evaluator Score	Evaluator Initials
1.	Develop and implement an effective initial action plan.			
2.	Analyze and use information gained in size-up.			
3.	Use resources in a reasonable, safe, and prudent manner.			
4.	Maintain supervision and accountability for personnel.			
5.	Communicate effectively using both verbal and written methods.			
6.	Implement and operate within the emergency management system.			
7.	Conduct a postincident analysis using proper policies, forms, and procedures.			

Evaluator: _____   _____

    Printed   **Signed**

Comments:

_____

_____

Candidate Name: _____   Date: _____

TASK SHEET: 4.7.1, 4.7.2 NFPA 1021, 2020 **General Requirements**	**Task:** Apply safety regulations at the unit level, given safety policies and procedures, so that required reports are completed, in-service training is conducted, and member responsibilities are conveyed. **Task:** Conduct an initial accident investigation, given incident and investigation forms, so that the incident is documented and reports are processed in accordance with policies and procedures.
PERFORMANCE OUTCOME:	Candidate will assume the role of fire department company officer and will conduct an initial accident investigation involving a fire department vehicle or injury. Provide an actual or simulated accident scenario. Candidate will interview witnesses, complete required reports, make recommendations on preventing future similar accidents, and convey responsibility for the accident to the appropriate person. Candidate will identify safety hazards or unsafe behaviors that may have contributed to the accident.

EQUIPMENT REQUIRED: Safety and investigative policies and procedures. Persons to act as witnesses to the incident for the candidate to interview. Applicable incident, investigation, and accident reports or forms. Actual or simulated accident scenario provided with photographs, sketches, circumstances, or witness statements to be presented as the candidate performs the investigation.

CONDITIONS: The candidate will complete all elements of the assigned task. Include candidate's narrative on task completion, forms, photos/drawings, etc., department policy or procedure.

No.	Task Steps	Date	Evaluator Score	Evaluator Initials
1.	Freeze apparatus in position to conduct investigation, if possible.			
2.	Make appropriate notifications according to policy.			
3.	Use all available resources to document incident and conditions.			
4.	Interview witnesses to obtain facts, if possible.			
5.	Identify factors contributing to the accident.			
6.	Complete appropriate forms, reports, and statements that are required policy.			

Evaluator: _____   _____
                       Printed   **Signed**

Comments:

_____

_____

Candidate Name: _____ Date: _____

TASK SHEET: 4.7.3 NFPA 1021, 2020 General Requirements	**Task:** Explain the benefits of being physically and medically capable of performing assigned duties and effectively functioning during peak physical demand activities, given current fire service trends and agency policies, so that the need to participate in wellness and fitness programs is explained to members.
PERFORMANCE OUTCOME:	Candidate will complete a case study on the national death and injuries documented in the fire service and how fire service safety and wellness initiatives can help prevent these issues. Show examples of how the organization is improving this issue and what improvements could be made to current programs in the organization. Then the candidate will present this case study to personnel in the organization.

EQUIPMENT REQUIRED: Access to national death and injuries information or related documents. Paper, pen/pencil, computer, if applicable.

CONDITIONS: The candidate will complete all elements of the assigned task. Include candidate's case study and documentation of presentation to the organization.

No.	Task Steps	Date	Evaluator Score	Evaluator Initials
1.	Identify the issues causing death and injuries in the fire service.			
2.	Establish fire service safety and wellness initiatives.			
3.	Demonstrate the ability to communicate in writing.			
4.	Demonstrate the ability to effectively communicate verbally.			

Evaluator: _____ _____

              Printed   **Signed**

Comments:

_____

_____

© Glen E. Ellman

Fire Officer II
Task Booklet

Candidate's Name _____
First   Middle   Last

Candidate's Address _____
Street   City   State   Zip Code

Candidate's Phone Number: _____

Candidate's Email Address: _____

Candidate's Identification #: _____

Department Affiliation: _____

Date Fire Officer II Course Completion: _____

Date Task Booklet was Initiated: _____

Date Task Booklet was Completed: _____

Certification Date: _____

Evaluator:

Signature of Evaluator _____

Printed Name of Evaluator _____

Date: _____

Evaluator's Relevant Qualification (or agency certification)

_____
_____
_____
_____
_____
_____
_____
_____
_____
_____

19

Candidate Name: _____   Date: _____

TASK SHEET: 5.2.1 NFPA 1021, 2020 General Requirements	**Task:** Initiate actions to maximize member performance and/or to correct unacceptable performance, given human resource policies and procedures, so that member and/or unit performance improves or the issue is referred to the next level of supervision. (a) Requisite knowledge: Human resource policies and procedures, problem identification, organizational behavior, group dynamics, leadership styles, types of power, and interpersonal dynamics. (b) Requisite skills: The ability to communicate orally and in writing, to solve problems, to increase teamwork, and to counsel members.
PERFORMANCE OUTCOME:	The candidate will assume the role of company officer supervising four fire fighters at a fire station. One of the fire fighters, a new father, was late twice last month. Each time, he has provided a reasonable cause for his tardiness and called in before the shift began to alert the company. No official action has yet been taken. Today, the fire fighter was late again and received no corrective action. The action taken is entirely up to the candidate but is required to correct unacceptable performance so that performance improves or the issue is referred to the next officer in the chain of command. Actions taken must be reasonable, defensible, and in accordance with human resources policies and procedures. Candidate will inform the fire fighter of the action taken and make a written report for purposes of documentation.

EQUIPMENT REQUIRED: Member to act as subordinate fire fighter. Apply human resource policies and procedures. Paper, pen/pencil, computer, if applicable.

CONDITIONS: The candidate will complete all elements of the assigned task.

No.	Task Steps	Date	Evaluator Score	Evaluator Initials
1.	Adequately describe to the fire fighter the nature of the problem.			
2.	Make it clear in plain language what level of performance is expected.			
3.	Choose an action designed to correct unacceptable performance.			
4.	Inform the fire fighter of the corrective action to be taken.			
5.	Follow human resources policies, procedures, or guidelines.			
6.	Complete a written report documenting the problem and action taken.			

Evaluator: _____   _____
                                    Printed   **Signed**

Comments:

_____

_____

Candidate Name: _____   Date: _____

TASK SHEET: 5.2.2 NFPA 1021, 2020 **General Requirements**	**Task:** Evaluate the job performance of assigned members, given personnel records and evaluation forms, so that each member's performance is evaluated accurately and reported according to human resource policies and procedures. (a) Requisite knowledge: Human resource policies and procedures, job descriptions, objectives of a member evaluation program, and common errors in evaluating. (b) Requisite skills: The ability to communicate orally and in writing and to plan and conduct evaluations.
PERFORMANCE OUTCOME:	The candidate will assume the role of a company officer conducting a job performance evaluation of an assigned subordinate member. Using department and human resource policies and procedures, personnel records/forms, and job description, conduct a performance evaluation interview and make a written report.

EQUIPMENT REQUIRED: Person (classmates, volunteers, others) to act as subordinate member. Job description, personnel records/forms, departmental and human resource policies and procedures. Paper, pen/pencil, computer, if applicable.

CONDITIONS: The candidate will complete all elements of the assigned task.

No.	Task Steps	Date	Evaluator Score	Evaluator Initials
1.	Gather all available performance information prior to evaluating.			
2.	Follow applicable policies/procedures and maintain privacy.			
3.	Measure employee performance against the written job description.			
4.	Plan the evaluation interview as a tool to enhance performance.			
5.	Make a written report of performance on proper form/record.			
6.	Use positive rather than negative reinforcement whenever possible.			

Evaluator: _____   _____
                            Printed   **Signed**

Comments:

_____

_____

Candidate Name: _____   Date: _____

TASK SHEET: 5.2.3 NFPA 1021, 2020 General Requirements	**Task:** Create a professional development plan for a member of the organization, given the requirements for promotion, so that the individual acquires the necessary knowledge, skills, and abilities to be eligible for the examination for the position.
	(a) Requisite knowledge: Development of a professional development guide, including mentoring sessions and job shadowing. (b) Requisite skills: The ability to communicate orally and in writing.

PERFORMANCE OUTCOME:	The candidate will create a written career development plan for a subordinate, which outlines the necessary knowledge, skills, abilities, and certifications that must be obtained to become eligible and prepared for a promotion (to engineer, lieutenant, etc.). The written career development plan shall include timelines for meeting milestones and set mentoring and job shadowing guidelines. The written career development plan shall be presented to the subordinate, discussed, and implemented.

EQUIPMENT REQUIRED: Person (classmates, volunteers, others) to act as subordinate member. Job description, departmental and human resource policies and procedures. Paper, pen/pencil, computer, if applicable.

CONDITIONS: The candidate will complete all elements of the assigned task.

No.	Task Steps	Date	Evaluator Score	Evaluator Initials
1.	Gather information about the job requirements for the promotional position.			
2.	Determine, with the subordinate, future goals, plans, wishes, etc.			
3.	Develop a written career development plan.			
4.	Ensure plan includes timelines and milestones of development.			
5.	Establish mentoring and job shadowing guidelines.			
6.	Present the written career development plan to the subordinate.			
7.	Implement the written career development plan.			
8.	Ensure the written career development plan is realistic and reflects the promotional prerequisites.			

Evaluator: _____   _____
                                    Printed   **Signed**

Comments:

_____

_____

Candidate Name: _____   Date: _____

TASK SHEET: 5.3.1 NFPA 1021. 2020 **General Requirements**	**Task:** Supervise multi-unit implementation of a community risk reduction program (CPR), provided with with an AHJ CRR plan, policies, and procedures to meet community needs. (a) Requisite Knowledge: Community demographics and service organizations, verbal and nonverbal communication, and the role and mission of the department and its CRR plan. (b) Requisite Skills: Familiarity with public relations and the ability to supervise and communicate.
PERFORMANCE OUTCOME:	The candidate shall develop a written proposal to supervise multi-unit implementation of a community risk reduction program (CPR) and provided with an AHJ CRR plan, policies, and procedures to meet community needs. The plan shall be presented to senior officers.

EQUIPMENT REQUIRED: The candidate will utilize an AHJ-specific issue that is provided to them by the training officer or senior officer. Paper and pen/pencil. Computer if applicable.

CONDITIONS: The candidate will complete all elements of the assigned task.

No.	Task Steps	Date	Evaluator Score	Evaluator Initials
1.	Use effective problem-solving methods.			
2.	Establish the need for a community risk reduction program.			
3.	Create a written proposal outlining the issue and the benefits involved.			
4.	Utilize effective format for proposal writing.			
5.	Effectively present the proposal to senior officers.			
6.	Describe the costs and benefits of the proposed program.			

Evaluator: _____   _____
                                 Printed   **Signed**

Comments:

_____

_____

Candidate Name: _____   Date: _____

TASK SHEET: 5.3.2
 	
NFPA 1021, 2020
 	
General Requirements
	Task: Explain the benefits to the organization of cooperating with allied organizations, given a specific problem or issue in the community, so that the purpose for establishing external agency relationships is clearly explained.
	
	  (a) Requisite knowledge: Understanding of the agency mission and goals, and the type and functions of external agencies in the community.
	  (b) Requisite skills: The ability to develop interpersonal relationships through oral and written communications.

PERFORMANCE OUTCOME:	The candidate shall develop a written proposal to implement an interagency program with an allied organization that identifies and addresses a specific problem or issue within the community, and how it affects the missions and goals of both agencies. The plan shall be presented to senior officers.

EQUIPMENT REQUIRED: The candidate will use an AHJ-specific issue that is provided to them by the training officer or senior officer. Paper, pen/pencil, computer, if applicable.

CONDITIONS: The candidate will complete all elements of the assigned task.

No.	Task Steps	Date	Evaluator Score	Evaluator Initials
1.	Use effective problem-solving methods.			
2.	Establish the need for an interagency program.			
3.	Create a written proposal outlining the issue and the benefits involved.			
4.	Use effective format for proposal writing.			
5.	Effectively present the proposal to senior officers.			
6.	Describe the costs and benefits of the proposed program.			

Evaluator: _____   _____
                           Printed   **Signed**

Comments:

_____

_____

Candidate Name: _____   Date: _____

TASK SHEET: 5.4.1 **NFPA 1021, 2020** **General Requirements**	**Task:** Develop a policy or procedure, given an assignment, so that the recommended policy or procedure identifies the problem and proposes a solution. (a) Requisite knowledge: Policies and procedures and problem identification. (b) Requisite skills: The ability to communicate in writing and to solve problems.
PERFORMANCE OUTCOME:	The candidate will create a written document containing a recommendation to senior officer(s). Given an existing problem, propose a change to a policy or procedure in accordance with departmental goals to solve a problem.

EQUIPMENT REQUIRED: Description of existing problem. Paper, pen/pencil, computer, if applicable.

CONDITIONS: The candidate will complete all elements of the assigned task.

No.	Task Steps	Date	Evaluator Score	Evaluator Initials
1.	Use effective problem-solving methods.			
2.	Make a written proposal to senior officer(s).			
3.	Establish the need for policy or procedure.			
4.	Direct the written proposal to the appropriate person(s).			
5.	Use effective format for proposal writing.			
6.	Describe cost and benefits of proposed change.			

Evaluator: _____   _____
                                Printed   **Signed**

Comments:

_____

_____

Candidate Name: _____   Date: _____

TASK SHEET: 5.4.2 NFPA 1021, 2020 **General Requirements**	**Task:** Develop a project or divisional budget, given schedules and guidelines concerning its preparation, so that capital, operating, and personnel costs are determined and justified. (a) Requisite knowledge: The supplies and equipment necessary for ongoing or new projects, repairs to existing facilities, new equipment, apparatus maintenance, personnel costs, appropriate budgeting system. (b) Requisite skills: The ability to allocate finances, to relate interpersonally, and to communicate orally and in writing.
PERFORMANCE OUTCOME:	The candidate will prepare a budget in the proper format accompanied by supporting data for a department project. Candidate will use department records, policies, procedures, or guidelines to develop the project budget.

EQUIPMENT REQUIRED: Pen/pencil, computer, if applicable. Budget forms and potential revenue sources. Budget policies and procedures. Reference data to be gathered by candidate.

CONDITIONS: The candidate will complete all elements of the assigned task.

No.	Task Steps	Date	Evaluator Score	Evaluator Initials
1.	Make a written budget proposal for the appropriate person.			
2.	Allocate and account for all capital, operating, and personnel costs.			
3.	Use the correct type of budget for the project/department			
4.	Justify the budget, cost vs. benefit.			
5.	Use clear and concise written communication.			
6.	Follow the department's policies, procedures, or guidelines.			

Evaluator: _____   _____
              Printed   **Signed**

Comments:

_____

_____

Candidate Name: _____   Date: _____

TASK SHEET: 5.4.3 NFPA 1021, 2020 **General Requirements**	**Task:** Describe the process of purchasing, including soliciting and awarding bids, given established specifications, to ensure competitive bidding so that the needs of the organization are met within the applicable federal, state/provincial, and local laws and regulations. (a) Requisite knowledge: Purchasing laws, policies, and procedures. (b) Requisite skills: The ability to use evaluative methods and to communicate orally and in writing.
PERFORMANCE OUTCOME:	The candidate will describe the process of purchasing, including soliciting and awarding bids, for a predetermined product with established specifications. Candidate will ensure competitive bidding is used and the entire process is documented.

EQUIPMENT REQUIRED: Pen/pencil, computer, if applicable. Purchasing forms and purchasing policies and procedures. Reference data to be gathered by candidate.

CONDITIONS: The candidate will complete all elements of the assigned task.

No.	Task Steps	Date	Evaluator Score	Evaluator Initials
1.	Gather all applicable information before beginning.			
2.	Describe the process of soliciting for bids both verbally and in writing.			
3.	Describe the process of awarding bids both verbally and in writing.			
4.	Describe the process of purchasing both verbally and in writing.			
5.	Ensure competitive bidding is used.			
6.	Use clear and concise written communication.			

Evaluator: _____   _____
                            Printed   **Signed**

Comments:

_____

_____

Candidate Name: _____   Date: _____

TASK SHEET: 5.4.4	**Task:** Prepare a news release, given an event or topic, so that the information is accurate and formatted correctly.
NFPA 1021, 2020	(a) Requisite knowledge: Policies and procedures and the format used for news releases.
General Requirements	(b) Requisite skills: The ability to communicate orally and in writing.

PERFORMANCE OUTCOME:	The candidate will prepare a news release for a specific event or topic. Candidate will use proper format and communicate the message clearly and accurately.

EQUIPMENT REQUIRED: News release policies and procedures. Event or topic. Pen/pencil, paper, computer, if applicable.

CONDITIONS: The candidate will complete all elements of the assigned task.

No.	Task Steps	Date	Evaluator Score	Evaluator Initials
1.	Create a written news release.			
2.	Gather all applicable information before beginning.			
3.	Use proper news release format.			
4.	Obey applicable policies and procedures.			
5.	Communicate effectively in writing.			
6.	Produce and deliver a clear and effective message.			

Evaluator: _____   _____
                                Printed   **Signed**

Comments:

_____

_____

Candidate Name: _____ Date: _____

TASK SHEET: 5.4.5 NFPA 1021, 2020 General Requirements	**Task:** Prepare a concise report for transmittal to a supervisor, given fire department record(s) and a specific request for details such as trends, variances, or other related topics so that the information required for the AHJ is accurate and documented. (a) Requisite knowledge: The data processing system. (b) Requisite skills: The ability to communicate in writing and to interpret data.
PERFORMANCE OUTCOME:	The candidate will answer a specific request for information regarding trends, variances, or other related topics from a supervisor. Candidate will use department records from which to gather information to create a written report to transmit to the supervisor.

EQUIPMENT REQUIRED: Specific request for information from a supervisor. Fire department records, information management system, or data processing system. Paper, pen/pencil, computer, if applicable.

CONDITIONS: The candidate will complete all elements of the assigned task.

No.	Task Steps	Date	Evaluator Score	Evaluator Initials
1.	Create a written report for transmittal to the supervisor.			
2.	Directly answer the specific request for information.			
3.	Use an appropriate report format.			
4.	Use clear and concise written communication.			
5.	Properly access reference data.			
6.	Correctly analyze and interpret reference data.			

Evaluator: _____   _____
                              Printed   **Signed**

Comments:
_____
_____

Candidate Name: _____   Date: _____

TASK SHEET: 5.4.6 NFPA 1021, 2020 **General Requirements**	**Task:** Develop a plan to accomplish change in the organization, given an agency's change of policy or procedures, so that the effective change is implemented in a positive manner. (a) Requisite knowledge: Planning and implementing change. (b) Requisite skills: The ability to clearly communicate orally and in writing.
PERFORMANCE OUTCOME:	Given a newly approved policy, SOP, or procedure, the candidate shall plan and implement the change within the agency. The change will reflect the intent of management and have the least intrusive impact as possible.

EQUIPMENT REQUIRED: The candidate will be provided a new policy, SOP, or procedure by the training officer or senior officer. Paper, pen/pencil, computer, if applicable.

CONDITIONS: The candidate will complete all elements of the assigned task.

No.	Task Steps	Date	Evaluator Score	Evaluator Initials
1.	Gather all applicable information.			
2.	Create a written plan for implementation of the change.			
3.	Disseminate and reinforce the need for the change to applicable personnel.			
4.	Implement the change.			
5.	Ensure training and documentation of acknowledgment by all affected personnel.			
6.	Follow agency procedures.			

Evaluator: _____   _____
                              Printed   **Signed**

Comments:

_____

_____

Candidate Name: _____   Date: _____

TASK SHEET: 5.5.1 NFPA 1021, 2020 **General Requirements**	**Task:** Determine the point of origin and preliminary cause of a fire, given a fire scene, photographs, diagrams, and pertinent data and/or sketches, to determine if arson is suspected so that law enforcement action is taken. (a) Requisite knowledge: Methods used by arsonists, common causes of fire, basic cause and origin determination, fire growth and development, and documentation of preliminary fire investigative procedures. (b) Requisite skills: The ability to communicate orally and in writing and to apply knowledge using deductive skills.
PERFORMANCE OUTCOME:	The candidate will be given a real or simulated fire incident scene. The candidate will determine the point of origin and identify a preliminary fire cause, using photographs, diagrams, pertinent data, and/or sketches. Candidate will determine if arson is suspected. Candidate will document preliminary investigation procedures and results.

EQUIPMENT REQUIRED: Real or simulated fire incident scene with materials necessary to create the proper environment. Applicable reports or witness statement forms. Photographs, diagrams, pertinent data, and/or sketches. Paper, pen/pencil, computer, if applicable.

CONDITIONS: The candidate will complete all elements of the assigned task.

No.	Task Steps	Date	Evaluator Score	Evaluator Initials
1.	Determine point of origin.			
2.	Identify a preliminary cause of the fire.			
3.	Use all sources of incident information available.			
4.	Use appropriate investigation techniques.			
5.	Document the procedure and results of preliminary investigation.			
6.	Include all pertinent data with the preliminary investigation report.			

Evaluator: _____   _____
                                    Printed   **Signed**

Comments:

_____

_____

Candidate Name: _____ Date: _____

TASK SHEET: 5.6.1 NFPA 1021, 2020 General Requirements	**Task:** Produce operational plans, given an emergency incident requiring multi-unit operations, so that required resources and the assignments are obtained and plans are carried out in compliance with approved safety procedures resulting in the mitigation of the incident.

<table>
<tr><td></td><td>(a) Requisite knowledge: Standard operating procedures; national, state/provincial, and local information resources available for the mitigation of emergency incidents; an incident management system; and a personnel accountability system.
(b) Requisite skills: The ability to implement an incident management system, to communicate orally, to supervise and account for assigned personnel under emergency conditions, and to serve in command staff and unit supervision positions within the incident management system.</td></tr>
</table>

PERFORMANCE OUTCOME:	The candidate will develop and implement an operational plan for hazardous materials incident scenario and another multi-unit emergency scenario. Candidate must analyze emergency scene condition, allocate resources, communicate verbally and in writing, operate within an emergency management system, supervise, and account for assigned personnel so that resources are effectively and safely deployed to mitigate the situation.

EQUIPMENT REQUIRED: One (1) hazardous materials incident scenario and one (1) multi-unit emergency scenario including type of incident, size-up information, and assigned resources. Policies and procedures. Personnel accountability system components. Pen/pencil, paper, computer, if applicable.

CONDITIONS: The candidate will complete all elements of the assigned task.

No.	Task Steps	Date	Evaluator Score	Evaluator Initials
1.	Identify the hazards and dangers associated with this hazardous material.			
2.	Describe the goal in managing this incident and the objectives to be completed in the next 8- to12-hour work period.			
3.	Implement necessary safety precautions and personnel accountability.			
4.	Produce an effective strategic operational plan to respond to this hazardous materials incident for the next 8- to 12-hour work period.			
5.	Allocate, supervise, and account for human and equipment resources at the task level.			
6.	Identify additional resources needed to meet the incident goal in the next 24 to 72 hours.			

Evaluator: _____   _____
                                 Printed   **Signed**

Comments:

_____

_____

Candidate Name: _____   Date: _____

TASK SHEET: 5.6.2 **NFPA 1021, 2020** **General Requirements**	**Task:** Develop and conduct a postincident analysis, given multi-unit incident and postincident analysis policies, procedures, and forms, so that all required critical elements are identified and communicated and the approved forms are completed and processed. (a) Requisite knowledge: Elements of a postincident analysis, basic building construction, basic fire protection systems and features, basic water supply, basic fuel loading, fire growth and development, and departmental procedures relating to dispatch response, strategy tactics and operations, and customer service. (b) Requisite skills: The ability to write reports, communicate orally, and evaluate skills.
PERFORMANCE OUTCOME:	Candidate will conduct a postincident analysis, given a multi-unit incident scenario. Candidate must be able to analyze the elements of a postincident analysis, identify all of the required critical elements, complete approved forms, and communicate verbally and in writing of the findings.

EQUIPMENT REQUIRED: Emergency multi-unit incident scenario including type of incident, size-up information, and assigned resources. Policies and procedures. Necessary forms and reports. Personnel accountability system components. Pen/pencil, paper, computer, if applicable.

CONDITIONS: The candidate will complete all elements of the assigned task.

No.	Task Steps	Date	Evaluator Score	Evaluator Initials
1.	Gather information from the multi-unit incident/scenario.			
2.	Analyze policies, procedures, guidelines, and forms.			
3.	Identify critical elements of a postincident analysis.			
4.	Complete approved forms.			
5.	Communicate effectively using both verbal and written methods.			

Evaluator: _____   _____
              Printed   **Signed**

Comments:
_____
_____

Candidate Name: _____ Date: _____

TASK SHEET: 5.6.3 NFPA 1021, 2020 **General Requirements**	**Task:** Prepare a written report, given incident reporting data from the jurisdiction, so that the major causes for service demands are identified for various planning areas within the service area of the organization. (a) Requisite knowledge: Analyzing data. (b) Requisite skills: The ability to write clearly and to interpret response data correctly to identify the reasons for service demands.
PERFORMANCE OUTCOME:	The candidate shall analyze the provided data and present a written report that summarizes the findings to a senior officer within the agency. The report must identify major causes for service demands within various planning areas within the jurisdiction.

EQUIPMENT REQUIRED: The candidate will be provided data from the agency records, information management system, data processing system, or incident reporting system. Paper, pen/pencil, computer, if applicable.

CONDITIONS: The candidate will complete all elements of the assigned task.

No.	Task Steps	Date	Evaluator Score	Evaluator Initials
1.	Analyze the data.			
2.	Determine the major causes for service demands within the planning area(s).			
3.	Prepare a written report outlining the major causes for service demands.			
4.	Effectively present the report to senior officer(s).			

Evaluator: _____   _____
                              Printed   **Signed**

Comments:

_____

_____

Candidate Name: _____   Date: _____

TASK SHEET: 5.7.1 NFPA 1021, 2020 General Requirements	**Task:** Analyze a member's accident, injury, or health exposure history, given a case study, so that a report is prepared for a supervisor and includes action taken and recommendations given.
	(a) Requisite knowledge: The causes of unsafe acts, health exposures, or conditions that result in accidents, injuries, occupational illnesses, or deaths. (b) Requisite skills: The ability to communicate in writing and to interpret accidents, injuries, occupational illnesses, or death reports.

PERFORMANCE OUTCOME:	The candidate will examine a case study of a member's accident injury or health exposure and prepare a written report for a supervisor. Report will identify unsafe environments and behaviors, document action taken, and make recommendations to prevent reoccurrence.

EQUIPMENT REQUIRED: Case Study described above. Health and safety policies and procedures. Injury/illness reports. Pen/pencil, paper, computer, if applicable.

CONDITIONS: The candidate will complete all elements of the assigned task.

No.	Task Steps	Date	Evaluator Score	Evaluator Initials
1.	Create a written report of illness, injury, or health exposure.			
2.	Include all contributing factors in the report based on the case study.			
3.	Identify unsafe work environment and/or behavior.			
4.	Document actions taken in response to illness, injury, or exposure.			
5.	Provide recommendations to prevent reoccurrence.			
6.	Present a clear and concise written report.			

Evaluator: _____   _____
                                    Printed   **Signed**

Comments:

_____

_____

Fire Officer III
Task Booklet

© Glen E. Ellman

Candidate's Name _____
          First   Middle   Last

Candidate's Address _____
          Street   City   State   Zip Code

Candidate's Phone Number: _____

Candidate's Email Address: _____

Candidate's Identification #: _____

Department Affiliation: _____

Date Fire Officer III Course Completion: _____

Date Task Booklet was Initiated: _____

Date Task Booklet was Completed: _____

Certification Date: _____

Evaluator:

Signature of Evaluator _____

Printed Name of Evaluator _____

Date: _____

Evaluator's Relevant Qualification (or agency certification)

_____

_____

_____

_____

_____

_____

_____

_____

_____

_____

Candidate Name: _____   Date: _____

TASK SHEET: 6.1.2, 6.2.1	**Task:** Establish minimum staffing requirements, given available human resources; policies and procedures; federal, state, and provincial laws; and rules and regulations, so that AHJ job-related credentials are maintained.
NFPA 1021, 2020	
General Requirements	

PERFORMANCE OUTCOME:	The candidate will establish personnel assignments in which the candidate relates interpersonally and communicates orally and in writing.

EQUIPMENT REQUIRED: Computer, pen, paper, department policy and procedure, department manpower levels.

CONDITIONS: Given knowledge, training, personnel roster, and experience of department members, the candidate shall:

No.	Task Steps	Date	Evaluator Score	Evaluator Initials
1.	Identify ways to maximize efficiency within the department.			
2.	Establish personnel assignments based on department efficiency study.			
3.	Develop organizational chart in accordance with department policy.			
4.	Communicate personnel assignments orally and in written format.			

Evaluator: _____   _____
                               Printed   **Signed**

Comments:

_____

_____

Candidate Name: _____   Date: _____

TASK SHEET: 6.1.2, 6.2.2 NFPA 1021, 2020 General Requirements	**Task:** Develop procedures for hiring members, given policies of the AHJ and legal requirements, so that the process is valid and reliable.

PERFORMANCE OUTCOME:	The candidate will create, revise, or establish procedures for hiring members in which the candidate relates interpersonally and communicates orally and in writing.

EQUIPMENT REQUIRED: Computer, pen, paper, department policies and procedures.

CONDITIONS: Given policies of the AHJ and legal requirements, the candidate shall:

No.	Task Steps	Date	Evaluator Score	Evaluator Initials
1.	Orally communicate with human resources on identifying proper laws, regulations, policies, and procedures related to human resource management.			
2.	Determine need for hiring members.			
3.	Develop or revise a job description and job announcement.			
4.	Develop or revise methods for recruiting potential new members.			
5.	Develop or revise the application screening process.			
6.	Develop or revise an examination process for hiring members that is valid and reliable.			
7.	Develop or revise a procedure that determines the hiring member's ability to meet the physical requirements as established in the job description.			
8.	Develop or revise an interview process for hiring members that is valid and reliable.			
9.	Establish or revise and submit criteria for candidate selection based on the hiring process.			

Evaluator: _____   _____
                                    Printed   **Signed**

Comments:

_____

_____

Candidate Name: _____   Date: _____

<table>
<tr>
<td colspan="2">TASK SHEET: 6.2.3

NFPA 1021, 2020

General Requirements</td>
<td colspan="5">Task: Develop procedures and programs for promoting members, given applicable policies and legal requirements, so that the process is valid and reliable, job-related, and nondiscriminatory.</td>
</tr>
<tr>
<td colspan="2" align="center">PERFORMANCE OUTCOME:</td>
<td colspan="5">The candidate will develop procedures and programs for promoting members in which the candidate relates interpersonally and communicates orally and in writing.</td>
</tr>
<tr>
<td colspan="7">EQUIPMENT REQUIRED: Computer, pen, paper, department policies and procedures.</td>
</tr>
<tr>
<td colspan="7">CONDITIONS: Given knowledge, training, and experience of department members, the candidate shall:</td>
</tr>
<tr>
<td>No.</td>
<td colspan="3" align="center">Task Steps</td>
<td>Date</td>
<td>Evaluator Score</td>
<td>Evaluator Initials</td>
</tr>
<tr>
<td>1.</td>
<td colspan="3">Create dialog with appropriate personnel for establishing job requirements for promotable positions.</td>
<td></td>
<td></td>
<td></td>
</tr>
<tr>
<td>2.</td>
<td colspan="3">Develop a job announcement for promotional opportunity.</td>
<td></td>
<td></td>
<td></td>
</tr>
<tr>
<td>3.</td>
<td colspan="3">Develop an application screening process.</td>
<td></td>
<td></td>
<td></td>
</tr>
<tr>
<td>4.</td>
<td colspan="3">Develop an examination process that validates member's ability to perform in the promotable position.</td>
<td></td>
<td></td>
<td></td>
</tr>
<tr>
<td>5.</td>
<td colspan="3">Develop an interview process for promoting members that is valid and reliable.</td>
<td></td>
<td></td>
<td></td>
</tr>
<tr>
<td>6.</td>
<td colspan="3">Establish criteria for candidate selection based on the promotion process.</td>
<td></td>
<td></td>
<td></td>
</tr>
</table>

Evaluator: _____   _____

   Printed   **Signed**

Comments:

_____

_____

Candidate Name: _____ Date: _____

TASK SHEET: 6.1.2, 6.2.4 NFPA 1021, 2020 General Requirements	**Task:** Describe methods to facilitate and encourage members to participate in professional development given a professional development model, so that members achieve their personal and professional goals.
PERFORMANCE OUTCOME:	The candidate will describe methods to encourage members to participate in professional development in which the candidate evaluates potential, communicates orally, and counsels members.

EQUIPMENT REQUIRED: Computer, pen, paper, department policies and procedures, department employee development plan.

CONDITIONS: Given knowledge, training, and experience of department members, the candidate shall:

No.	Task Steps	Date	Evaluator Score	Evaluator Initials
1.	Apply interpersonal skills to facilitate a discussion with members on ways to enhance their professional development.			
2.	Use appropriate written and verbal counseling and motivational skills to encourage members to participate in professional development.			
3.	Develop written documentation that outlines and tracks professional development progress.			

Evaluator: _____ _____
                      Printed   **Signed**

Comments:

_____

_____

Candidate Name: _____   Date: _____

TASK SHEET: 6.2.5 NFPA 1021, 2020 General Requirements	**Task:** Develop a proposal for improving an employee benefit, given a need in the organization, so that adequate information is included to justify the requested benefit improvement.
PERFORMANCE OUTCOME:	The candidate shall develop a proposal for improving an employee benefit in which the candidate conducts research and communicates orally and in writing.

EQUIPMENT REQUIRED: Computer, pen, paper, department policies and procedures.

CONDITIONS: Given a department benefit program, the candidate shall:

No.	Task Steps	Date	Evaluator Score	Evaluator Initials
1.	Determine need within the organization for improving an employee benefit via appropriate dialog with department personnel.			
2.	Establish a funding source for the employee benefit that is being developed or improved.			
3.	Develop a proposal with adequate information to justify the requested benefit improvement.			
4.	Provide appropriate documentation to organization members on the approval or denial for the improvement of the requested benefit.			

Evaluator: _____   _____
                                Printed   **Signed**

Comments:

_____

_____

Candidate Name: _____   Date: _____

TASK SHEET: 6.1.2, 6.2.6 NFPA 1021, 2020 **General Requirements**	**Task:** Develop a plan for providing an employee accommodation, given the need, as well as the requirements and applicable law, so that adequate information is included to justify the requested change(s).

PERFORMANCE OUTCOME:	The candidate shall develop a plan for providing an employee accommodation in which the candidate conducts research and communicates orally and in writing.

EQUIPMENT REQUIRED: Computer, pen, paper, department policies and procedures.

CONDITIONS: Given legal requirements and organization policies and procedures, the candidate shall:

No.	Task Steps	Date	Evaluator Score	Evaluator Initials
1.	Research laws, regulations, policies, and procedures to ensure current accommodations comply with federal employee accommodation requirements.			
2.	Develop a plan for providing employee accommodations should the conducted research indicate that the organization does not comply with federal employee accommodation requirements.			
3.	Disseminate acquired information on appropriate employee accommodations to members of the organization both orally and in written form.			

Evaluator: _____   _____
                                 Printed   **Signed**

Comments:

_____

_____

Candidate Name: _____   Date: _____

TASK SHEET: 6.2.7 NFPA 1021, 2020 General Requirements	**Task:** Develop an ongoing education training program, given organizational training requirements, so that members of the organization are given appropriate training to meet the mission of the organization.

PERFORMANCE OUTCOME:	The candidate shall provide a needs assessment in which the candidate relates interpersonally and communicates orally and in writing.

EQUIPMENT REQUIRED: Computer, pen, paper, department policies and procedures.

CONDITIONS: Given department training policies, procedures, and/or manuals, the candidate shall:

No.	Task Steps	Date	Evaluator Score	Evaluator Initials
1.	Conduct a needs assessment for the development of an ongoing education training program.			
2.	Provide justification for current agency policies and procedures for ongoing education.			
3.	Provide an improvement plan for current training practices based on a needs assessment.			
4.	Conduct a formal briefing to Command staff in reference to needs assessment, plan development, and implementation of the new ongoing education training program.			

Evaluator: _____   _____
                                    Printed   **Signed**

Comments:

_____

_____

Candidate Name: _____   Date: _____

TASK SHEET: 6.3.1 NFPA 1021, 2020 General Requirements	**Task:** Develop a community risk reduction program, given risk assessment data, so that program outcomes are met.
PERFORMANCE OUTCOME:	The candidate shall prepare a community risk reduction program in which the candidate relates interpersonally and communicates orally and in writing.

EQUIPMENT REQUIRED: Computer, pen, paper, department policies and procedures.

CONDITIONS: Given current community demographics and needs, the candidate shall:

No.	Task Steps	Date	Evaluator Score	Evaluator Initials
1.	Define customer service principles tied to the agency's mission or values statement.			
2.	Define current service responsibilities.			
3.	Identify both internal and external resources required for the community risk reduction program.			
4.	Provide justification for the program based on a needs assessment that identifies community demographics and unmet needs.			
5.	Prepare a community risk reduction program that contains: Approved goals and objectives related to safety, injury prevention, and convenient public services An expected outcome Method for introduction to the public Method for monitoring and maintaining the program Identification of person(s) responsible for program management Method for tracking results or trends from the program.			
6.	Verbally deliver the prepared community risk reduction program to department personnel and external customers.			

Evaluator: _____   _____
                              Printed   **Signed**

Comments:
_____
_____

Candidate Name: _____ Date: _____

TASK SHEET: 6.4.1 NFPA 1021, 2020 General Requirements	**Task:** Develop a divisional or departmental budget, given schedules and guidelines concerning its preparation, so that capital, operating, and personnel costs are determined and justified.

PERFORMANCE OUTCOME:	The candidate shall develop a divisional or departmental budget that demonstrates an ability to allocate finances, relate interpersonally, and communicate orally and in writing.

EQUIPMENT REQUIRED: Computer, pen, paper, department policies and procedures.

CONDITIONS: Given department schedules, policies, procedures, and guidelines, the candidate shall:

No.	Task Steps	Date	Evaluator Score	Evaluator Initials
1.	Describe internal finance policies and procedures.			
2.	Identify internal systems that monitor fiscal resources.			
3.	Develop a budget that addresses the needs of current and new programs.			
4.	Highlight portions of a budget that address a need for new equipment, apparatus and facility maintenance, and personnel costs.			
5.	Communicate the budget management system both verbally and in writing to appropriate personnel.			

Evaluator: _____   _____
                        Printed   **Signed**

Comments:

_____

_____

Candidate Name: _____   Date: _____

TASK SHEET: 6.4.2 NFPA 1021, 2020 General Requirements	**Task:** Develop a budget management system, given fiscal and financial policies, so that the division or department stays within the budgetary authority.

PERFORMANCE OUTCOME:	The candidate shall develop a budget management system in which the candidate interprets financial data and communicates orally and in writing.

EQUIPMENT REQUIRED: Computer, pen, paper, department policies and procedures.

CONDITIONS: Given department schedules, policies, procedures, and guidelines, the candidate shall:

No.	Task Steps	Date	Evaluator Score	Evaluator Initials
1.	Describe internal finance policies and procedures.			
2.	Identify internal systems that monitor fiscal resources.			
3.	Develop a budget management system that tracks accounts payable and receivable.			
4.	Ensure the budget management system meets recommended financial requirements for internal and external auditing.			
5.	Communicate the budget management system both verbally and in writing to appropriate personnel.			

Evaluator: _____   _____
               Printed   **Signed**

Comments:

_____

_____

Candidate Name: _____ Date: _____

TASK SHEET: 6.4.3 NFPA 1021, 2020 General Requirements	**Task:** Describe the agency's process for developing requests for proposals (RFPs) and soliciting and awarding bids, given established specifications and the agency's policies and procedures, so that competitive bidding is ensured.
PERFORMANCE OUTCOME:	The candidate shall describe the agency's process for developing requests for proposals and soliciting and awarding bids in which the candidate relates interpersonally as well as communicates orally and in writing

EQUIPMENT REQUIRED: Computer, pen, paper, department policies and procedures.

CONDITIONS: Given established specifications, policies, and procedures, the candidate shall:

No.	Task Steps	Date	Evaluator Score	Evaluator Initials
1.	Describe the agency's process for developing requests for proposals.			
2.	Describe the agency's method for soliciting bids in which competitive bidding is ensured.			
3.	Describe the agency's method for awarding bids.			
4.	Establish a forum in which requests for proposals can be communicated both verbally and in writing to potential vendors.			

Evaluator: _____   _____
                                 Printed   **Signed**

Comments:

_____

_____

Candidate Name: _____ Date: _____

TASK SHEET: 6.4.4 NFPA 1021, 2020 General Requirements	**Task:** Direct the development, maintenance, and evaluation of a department record and management system, given policies and procedures, so that completeness and accuracy are achieved.
PERFORMANCE OUTCOME:	The candidate shall direct the development, maintenance, and evaluation of a department record and management system in which the candidate demonstrates the ability to use evaluative methods, organize data, and communicate orally and in writing.

EQUIPMENT REQUIRED: Computer, pen, paper, department policies and procedures.

CONDITIONS: Given department policies and procedures, the candidate shall:

No.	Task Steps	Date	Evaluator Score	Evaluator Initials
1.	Evaluate current policies and procedures for developing, maintaining, and evaluating department records.			
2.	Determine the need for developing or purchasing a department record and management system based on review of policies and procedures.			
3.	Establish funding source for the purchase or development of a department record and management system.			
4.	Organize data on various record and management systems available to fire departments.			
5.	Establish and communicate in verbal and written format maintenance guidelines for the system that has been purchased or developed.			
6.	Develop an evaluation method for determining effectiveness and accuracy of the records and management system.			

Evaluator: _____   _____
                                   Printed   **Signed**

Comments:

_____

_____

Candidate Name: _____   Date: _____

TASK SHEET: 6.1.2, 6.4.5 NFPA 1021, 2020 General Requirements	**Task:** Analyze and interpret records and data, given a fire department records system, so that validity is determined and improvements are recommended.

PERFORMANCE OUTCOME:	The candidate shall analyze and interpret records and data in which the candidate uses evaluative methods and communicates both orally and in writing.

EQUIPMENT REQUIRED: Computer, pen, paper, department policies and procedures.

CONDITIONS: Given department policies and procedures, the candidate shall:

No.	Task Steps	Date	Evaluator Score	Evaluator Initials
1.	Complete a written analysis of fire department records and data management system (i.e., current system accurately captures data, storage capabilities, and retrieval of department information).			
2.	Provide documentation of recommended improvements based on findings.			
3.	Determine a funding source for recommended improvements based on findings.			
4.	Verbally communicate and submit written analysis and documentation to appropriate department or city personnel.			

Evaluator: _____   _____
                                  Printed   **Signed**

Comments:

_____

_____

Candidate Name: _____   Date: _____

TASK SHEET: 6.1.2, 6.4.6 NFPA 1021, 2020 General Requirements	**Task:** Develop a model for continuous organizational improvement, given resources for an area to be protected, so that resource use is maximized.

PERFORMANCE OUTCOME:	The candidate shall develop a model for continuous organizational improvement in which the candidate demonstrates the ability to research, use evaluative methods, analyze data, and communicate both orally and in writing.

EQUIPMENT REQUIRED: Computer, pen, paper, department policies and procedures.

CONDITIONS: Given resources for an area to be protected and department policies and procedures, the candidate shall:

No.	Task Steps	Date	Evaluator Score	Evaluator Initials
1.	Provide/cite research community hazards and needs based on the following: Community risk analysis Community needs identification Community-required services Accreditation programs			
2.	Evaluate local risks and planning for necessary resource use.			
3.	Analyze gathered information for assistance in the development of the following: Organizing and deploying resources Leadership strategies for the political process Strategic planning at the department level.			
4.	Forecast what impact the model for continuous organizational improvement will have on the community as well as what problems may be encountered.			
5.	Develop and communicate a written plan motivating government representatives and administrative personnel in the organization to make improvements in the community.			

Evaluator: _____   _____
                         Printed   **Signed**

Comments:

_____

_____

Candidate Name: _____   Date: _____

TASK SHEET: 6.1.2, 6.5.1 NFPA 1021, 2020 General Requirements	**Task:** Evaluate the inspection program of the AHJ, given current program goals, objectives, performance data, and resources, so that the results are evaluated to determine effectiveness.
PERFORMANCE OUTCOME:	The candidate shall evaluate the inspection program of the AHJ to determine effectiveness in which the candidate demonstrates the ability to use evaluative methods, to analyze data, and to communicate orally and in writing.

EQUIPMENT REQUIRED: Computer, pen, paper, department policies and procedures.

CONDITIONS: Given current program goals, objectives, performance data, and resources, the candidate shall:

No.	Task Steps	Date	Evaluator Score	Evaluator Initials
1.	Review and validate current program goals, objectives, performance data, and resources.			
2.	Conduct a risk assessment based on fire safety code violations for the AHJ.			
3.	Determine if program goals and objectives are being met based on compliance and trends from businesses within the AHJ.			
4.	Provide and implement a plan that addresses code violations and trends within the AHJ.			
5.	Give a presentation to administrative personnel outlining the new plan and reasons for the plan.			

Evaluator: _____   _____
                              Printed   **Signed**

Comments:

_____

_____

Candidate Name: _____   Date: _____

TASK SHEET: 6.5.2 NFPA 1021, 2020 **General Requirements**	**Task:** Develop a plan, given an identified fire safety problem, so that the approval for a new program, piece of legislation, form of public education, or fire safety code is facilitated.
PERFORMANCE OUTCOME:	The candidate will demonstrate the ability to use evaluative methods, to use consensus-building techniques, to communicate orally and in writing, and to organize plans.

EQUIPMENT REQUIRED: Computer, pen, paper, department policies and procedures.

CONDITIONS: Given department policies, procedures, codes, and a reoccurring fire problem of exceeding occupancy load at a local business, the candidate shall:

No.	Task Steps	Date	Evaluator Score	Evaluator Initials
1.	Review gathered inspection data to determine trends of code violations or fire safety problems.			
2.	Develop a plan that addresses trends and provides recommendations for improvement or elimination of problem chosen. Circle type to be facilitated: NEW PROGRAM PIECE OF LEGISLATION FORM OF PUBLIC EDUCATION FIRE SAFETY CODE			
3.	Present written and verbal documentation for the method used to address the fire safety problem using consensus-building techniques.			

Evaluator: _____   _____
                                Printed   **Signed**

Comments:

_____

_____

Candidate Name: _____   Date: _____

TASK SHEET: 6.6.1 **NFPA 1021, 2020** **General Requirements**	**Task:** Prepare an action plan, given an emergency incident requiring multiple agency operations, so that the required resources are determined and the resources are assigned and placed to mitigate the incident.

PERFORMANCE OUTCOME:	The candidate shall prepare an action plan in which the candidate uses an evaluative method, delegates authority, organizes a plan, and communicates both orally and in writing.

EQUIPMENT REQUIRED: Computer, pen, paper, department policies and procedures.

CONDITIONS: Given a multi-agency emergency scenario, including type of incident, size-up information, assigned agencies, policies, and procedures, the candidate shall:

No.	Task Steps	Date	Evaluator Score	Evaluator Initials
1.	Prepare an Incident Action Plan to mitigate the multi-agency emergency incident based on an effective evaluation of the incident.			
2.	Allocate, supervise, and account for human and equipment resources.			
3.	Implement necessary safety precautions and personnel accountability.			
4.	Verbally communicate and provide completed Incident Action Plan to appropriate personnel during briefing.			

Evaluator: _____   _____
                         Printed   **Signed**

Comments:

_____

_____

Candidate Name: _____   Date: _____

TASK SHEET: 6.6.2	**Task:** Develop and conduct a postincident analysis, given multi-agency incident and postincident analysis policies, procedures, and forms, so that all required critical elements are identified and communicated and the appropriate forms are completed and processed in accordance with policies and procedures.
NFPA 1021, 2020	
General Requirements	

PERFORMANCE OUTCOME:	The candidate shall develop and conduct a postincident analysis in which the candidate demonstrates the ability to write reports, use evaluative skills, and communicate orally.

EQUIPMENT REQUIRED: Computer, pen, paper, department policies and procedures.

CONDITIONS: Given a multi-agency incident or scenario, postincident analysis policies, procedures, and forms, the candidate shall:

No.	Task Steps	Date	Evaluator Score	Evaluator Initials
1.	Gather information from the multi-agency incident/scenario.			
2.	Analyze policies, procedures, guidelines, and forms.			
3.	Identify critical elements of a postincident analysis.			
4.	Complete approved forms.			
5.	Conduct a postincident analysis using both verbal and written methods that includes all agencies involved.			

Evaluator: _____   _____
                             Printed   **Signed**

Comments:

_____

_____

Candidate Name: _____   Date: _____

TASK SHEET: 6.6.3	**Task:** Develop a plan for the agency, given an unmet need for resources that exceed what is available in the organization, so that the mission of the organization is capable of being performed in times of extraordinary need.
NFPA 1021, 2020	
General Requirements	

PERFORMANCE OUTCOME:	The candidate shall develop a plan for unmet needs in which the candidate conducts a needs assessment, evaluates external resources, and develops a plan.

EQUIPMENT REQUIRED: Computer, pen, paper, department policies and procedures.

CONDITIONS: Given current department policies, procedures, and response capabilities, the candidate shall:

No.	Task Steps	Date	Evaluator Score	Evaluator Initials
1.	Conduct a needs assessment based on current internal and external resources for extraordinary events.			
2.	Evaluate mutual aid and auto-aid agreements for effectiveness in assisting the organization in meeting resource needs during extraordinary events.			
3.	Develop a plan that addresses unmet needs for resources both internally and externally for events that exceed department capabilities while meeting the mission of the organization.			

Evaluator: _____   _____
                        Printed   **Signed**

Comments:

_____

_____

Candidate Name: _____   Date: _____

TASK SHEET: 6.7.1 NFPA 1021, 2020 **General Requirements**	**Task:** Develop a measurable accident and injury prevention program, given relevant local and national data, so that the results are evaluated to determine effectiveness of the program.
PERFORMANCE OUTCOME:	The candidate shall develop a plan for unmet needs in which the candidate conducts a needs assessment, evaluates external resources, and develops a plan.

EQUIPMENT REQUIRED: Computer, pen, paper, department policies and procedures.

CONDITIONS: Given department policies and procedures as well as local and national data, the candidate shall:

No.	Task Steps	Date	Evaluator Score	Evaluator Initials
1.	Evaluate current policies and procedures for effectiveness of the department's accident and injury prevention program.			
2.	Develop a measurable accident and injury prevention program based on data analysis in the review process.			
3.	Ensure the program's validity through an established data collection system, which includes maintaining permanent records of all accidents, injuries, illnesses, or deaths related to duty assignments.			
4.	Establish a reporting system in which program effectiveness can be communicated to department members, both orally and in writing, without affecting personnel privacy.			

Evaluator: _____   _____
                                Printed   **Signed**

Comments:

_____

_____

Candidate Name: _____ Date: _____

TASK SHEET: 6.8.1 NFPA 1021, 2020 **General Requirements**	**Task:** Develop a plan for the integration of fire service resources in the community's emergency management plan, given the requirements of the community and the resources available in the fire department, so that the role of the fire service is in compliance with local, state/provincial, and national requirements.
PERFORMANCE OUTCOME:	The candidate shall develop a plan for the integration of fire services resources in the community's emergency management plan in which the candidate demonstrates familiarity with emergency management interagency planning and coordination while communicating orally and in writing.

EQUIPMENT REQUIRED: Computer, pen, paper, department policies and procedures.

CONDITIONS: Based on laws, regulations, policies, and procedures pertaining to local, state, and federal emergency operations plans, the candidate shall:

No.	Task Steps	Date	Evaluator Score	Evaluator Initials
1.	Identify and define the current roles and responsibilities of the agency during large-scale emergencies.			
2.	Identify potential shortfalls within the agency for training, equipment, exercises, or cooperative agreements to meet existing hazards within the response area.			
3.	Justify the agency's roles and responsibilities for disaster response and mitigation efforts or offer recommendations for future improvements.			
4.	Justify the current emergency support functions for the agency or develop improvements for the integration and use of fire resources with regional, state, and federal assets during a major disaster incident.			
5.	Disseminate both verbally and in written format the revised emergency operations plan to appropriate local, state, and federal resources, if applicable.			

Evaluator: _____   _____
                            Printed   **Signed**

Comments:

_____

_____

© Glen E. Ellman

Fire Officer IV
Task Booklet

Candidate's Name _____
             First   Middle   Last

Candidate's Address _____
             Street   City   State   Zip Code

Candidate's Phone Number: _____

Candidate's Email Address: _____

Candidate's Identification #: _____

Department Affiliation: _____

Date Fire Officer IV Course Completion: _____

Date Task Booklet was Initiated: _____

Date Task Booklet was Completed: _____

Certification Date: _____

Evaluator:

Signature of Evaluator _____

Printed Name of Evaluator _____

Date: _____

Evaluator's Relevant Qualification (or agency certification)

_____

_____

_____

_____

_____

_____

_____

_____

_____

Candidate Name: _____   Date: _____

TASK SHEET: 7.2.1 **NFPA 1021, 2020** **General Requirements**	**Task:** Appraise the department's human resource demographics, given appropriate community demographic data, so that the recruitment, selection, and placement of human resources is effective and consistent with law and current best practices.

PERFORMANCE OUTCOME:	The candidate will demonstrate the ability to communicate, relate interpersonally, delegate authority, analyze issues, and solve problems.

EQUIPMENT REQUIRED: Computer, pen, paper, department policies and procedures.

CONDITIONS: Given policies and procedures; local, state/provincial, and federal regulations; community demographics; community issues; and formal and informal community, the candidate shall:

No.	Task Steps	Date	Evaluator Score	Evaluator Initials
1.	Identify applicable laws and standards.			
2.	Identify methods for evaluating demographics and practices.			
3.	Identify community demographics.			
4.	Complete an evaluation of demographics and practices.			
5.	Identify SWOT.			
6.	Identify need for corrective action, as needed.			
7.	Identify time frame for action.			
8.	Identify time for re-evaluation.			
9.	Delegate authority as needed for resolution.			
10.	Complete an executive summary.			
11.	Present summary of findings.			

Evaluator: _____   _____
                                Printed   **Signed**

Comments:

_____

_____

Candidate Name: _____ Date: _____

TASK SHEET: 7.2.2 NFPA 1021, 2020 **General Requirements**	**Task:** Initiate the development of a program, given current member/management relations, so that a positive and participative member/management program exists.

PERFORMANCE OUTCOME:	The candidate will demonstrate the ability to communicate, negotiate, analyze current status of member relations, relate interpersonally, analyze the current member/management relations, and conduct program implementation.

EQUIPMENT REQUIRED: Computer, pen, paper, department policies and procedures.

CONDITIONS: Given policies and procedures; contractual agreements; and local, state/provincial, and federal regulations, the candidate shall:

No.	Task Steps	Date	Evaluator Score	Evaluator Initials
1.	Identify methods for evaluating employee/management relations.			
2.	Identify data sources for evaluation.			
3.	Assess data and facts.			
4.	Identify corrective action plan process.			
5.	Identify a time frame for corrective action and follow-up.			
6.	Compile and summarize findings.			
7.	Present a summary of findings.			

Evaluator: _____   _____
                                Printed   **Signed**

Comments:

_____

_____

Candidate Name: _____   Date: _____

TASK SHEET: 7.2.3 NFPA 1021, 2020 **General Requirements**	**Task:** Evaluate the organization's education and in-service training program, given a summary of the job requirements for all positions within the department, so that all members can achieve and maintain required proficiencies.
PERFORMANCE OUTCOME:	The candidate will demonstrate the ability to communicate and to analyze and organize data and resources.

EQUIPMENT REQUIRED: Computer, pen, paper, department policies and procedures.

CONDITIONS: Given training resources; community needs; internal and external customers; policies and procedures; contractual agreements; and local, state/provincial, and federal regulations, the candidate shall:

No.	Task Steps	Date	Evaluator Score	Evaluator Initials
1.	Complete a summary of all requirements by position.			
2.	Identify department training and education needs by type.			
3.	Identify a timeframe for required training.			
4.	Evaluate existing training resources and options.			
5.	Identify and evaluate the training budget.			
6.	Select and justify training program goals and objectives.			
7.	Evaluate and enumerate the steps to implement a training program.			
8.	Identify a plan for evaluating the program.			
9.	Present a summary of findings.			

Evaluator: _____   _____
                                   Printed   **Signed**

Comments:

_____

_____

Candidate Name: _____ Date: _____

TASK SHEET: 7.2.4 NFPA 1021, 2020 General Requirements	**Task:** Appraise the member-assistance program, given data, so that the program, when used, produces stated program outcomes.

PERFORMANCE OUTCOME:	The candidate will demonstrate the ability to communicate, to relate interpersonally to members, and to analyze needs and results.

EQUIPMENT REQUIRED: Computer, pen, paper, department policies and procedures.

CONDITIONS: Given policies and procedures; available assistance programs; contractual agreements; and local, state/provincial, and federal regulations, the candidate shall:

No.	Task Steps	Date	Evaluator Score	Evaluator Initials
1.	Identify an evaluation method used for appraisal.			
2.	Identify employee assistance program (EAP) use.			
3.	Provide a description of the EAP.			
4.	Identify goals of the EAP.			
5.	Provide a summary of the EAP, including data, findings, conclusion, and recommendation.			
6.	Present a summary of findings.			

Evaluator: _____   _____
                                        Printed   **Signed**

Comments:

_____

_____

Candidate Name: _____   Date: _____

TASK SHEET: 7.2.5 **NFPA 1021, 2020** **General Requirements**	**Task:** Evaluate an incentive program, given data, so that a determination is made regarding achievement of the desired results, and modify as necessary.

PERFORMANCE OUTCOME:	The candidate will demonstrate the ability to communicate, relate interpersonally to members, and analyze programs.

EQUIPMENT REQUIRED: Computer, pen, paper, department policies and procedures.

CONDITIONS: Given policies and procedures; available incentive programs; contractual agreements; and local, state/provincial, and federal regulations, the candidate shall:

No.	Task Steps	Date	Evaluator Score	Evaluator Initials
1.	Identify an evaluation method used for evaluating the incentive program.			
2.	Identify incentive program use.			
3.	Provide a description of the incentive program.			
4.	Identify the goals of the incentive program.			
5.	Provide a summary of the incentive program, including data, findings, conclusion, and recommendation.			
6.	Identify a timeframe to implement the incentive program.			
7.	Present a summary of findings.			

Evaluator: _____   _____
                            Printed   **Signed**

Comments:

_____

_____

Candidate Name: _____   Date: _____

TASK SHEET: 7.3.1 NFPA 1021, 2020 General Requirements	**Task:** Attend, participate in, and assume a leadership role in community functions, given community needs, so that the image of the organization is enhanced.
PERFORMANCE OUTCOME:	The candidate will demonstrate the familiarity with public relations and the ability to communicate.

EQUIPMENT REQUIRED: Computer, pen, paper, department policies and procedures.

CONDITIONS: Given community demographics and socioeconomics, community and civic issues, effective customer service methods, and formal and informal community leaders, the candidate shall:

No.	Task Steps	Date	Evaluator Score	Evaluator Initials
1.	Identify required roles in community leadership.			
2.	Assess current roles and determine necessary changes.			
3.	Identify appropriate data sources.			
4.	Identify departmental activities for participation in a leadership role.			
5.	Present findings.			

Evaluator: _____   _____
                                           Printed   **Signed**

Comments:

_____

_____

Candidate Name: _____   Date: _____

TASK SHEET: 7.3.2 **NFPA 1021, 2020** **General Requirements**	**Task:** Develop and administer a media relations program, given AHJ policies and procedures, so that the dissemination of information is accurate and accessible.
PERFORMANCE OUTCOME:	The candidate will demonstrate the techniques of public relations, the ability to communicate, and crisis management.

EQUIPMENT REQUIRED: Computer, pen, paper, department policies and procedures.

CONDITIONS: Given AHJ policies and procedures for information dissemination; applicable laws, rules, and regulations governing information release; fundamentals of media relations; and social media platforms, the candidate shall:

No.	Task Steps	Date	Evaluator Score	Evaluator Initials
1.	Identify his or her required role in community leadership and information dissemination.			
2.	Identify appropriate data sources.			
3.	Develop techniques for information dissemination.			
4.	Identify departmental activities for participation in media relations and social media.			
5.	Present findings.			

Evaluator: _____   _____
                                Printed   **Signed**

Comments:

_____

_____

Candidate Name: _____   Date: _____

TASK SHEET: 7.4.1 NFPA 1021, 2020 General Requirements	**Task:** Develop a comprehensive long-range plan, given community requirements, current department status, and resources, so that the projected needs of the community are met.

PERFORMANCE OUTCOME:	The candidate will demonstrate a familiarity with public relations and the ability to communicate.

EQUIPMENT REQUIRED: Computer, pen, paper, department policies and procedures.

CONDITIONS: Given policies and procedures; physical and geographic characteristics; demographics; community plan; staffing requirements; response time benchmarks; contractual agreements; and local, state/provincial, and federal regulations, the candidate shall:

No.	Task Steps	Date	Evaluator Score	Evaluator Initials
1.	Develop an executive summary/statement.			
2.	Review and analyze the data.			
3.	Identify the departmental needs.			
4.	Select and define goals and objectives.			
5.	Develop requirements for meeting goals and objectives for the department.			
6.	Summarize alternatives.			
7.	Implement the plan with time frames.			
8.	Identify budget requirements.			
9.	Evaluate periodically.			
10.	Present findings.			

Evaluator: _____   _____
                                 Printed   **Signed**

Comments:

_____

_____

Candidate Name: _____ Date: _____

TASK SHEET: 7.4.2	**Task:** Evaluate and forecast training requirements, facilities, and buildings' needs, given data that reflect community needs and resources, so that departmental training goals are met.
NFPA 1021, 2020	
General Requirements	

PERFORMANCE OUTCOME:	The candidate will demonstrate the ability to communicate, make public presentations, interpret fiscal analysis, forecast needs, and analyze data.

EQUIPMENT REQUIRED: Computer, pen, paper, department policies and procedures.

CONDITIONS: Given policies and procedures; physical and geographic characteristics; building and fire codes; departmental plan; staffing requirements; training standards; needs assessment; contractual agreements; and local, state/provincial, and federal regulations, the candidate shall:

No.	Task Steps	Date	Evaluator Score	Evaluator Initials
1.	Identify a method for evaluating training and facility needs.			
2.	Review and analyze data to identify departmental needs.			
3.	Select and define goals and objectives.			
4.	Develop requirements for meeting goals and objectives for the department			
5.	Summarize alternatives.			
6.	Implement the plan with timeframes.			
7.	Identify budget requirements.			
8.	Evaluate periodically.			
9.	Develop an executive summary/statement.			
10.	Present a summary of findings.			

Evaluator: _____ _____
            Printed   **Signed**

Comments:

_____

_____

Candidate Name: _____   Date: _____

TASK SHEET: 7.4.3 NFPA 1021, 2020 General Requirements	**Task:** Complete a written, comprehensive, all-hazard risk and value analysis of the community, given the appropriate features of the service area of the organization, so that an accurate evaluation is made for service delivery decision-making.
PERFORMANCE OUTCOME:	The candidate will demonstrate the ability to conduct a needs assessment plan, to effectively communicate in writing, and to problem solve.

EQUIPMENT REQUIRED: Computer, pen, paper, department policies and procedures.

CONDITIONS: Given risk, hazard, and value analysis methods and process, as well as community development features, community demographics, and assessed valuation of properties in the community, the candidate shall:

No.	Task Steps	Date	Evaluator Score	Evaluator Initials
1.	Identify department mission and operational direction.			
2.	Identify data sources to determine comprehensive risks, trends, and issues for analysis.			
3.	Identify (when applicable) potential hazards and consequences resulting from inadequate attention to: a. accident and/or injury prevention programs and investigations. b. facility, apparatus, equipment, and personal protective equipment (PPE) inspection practices. c. personnel health and well-being policies and practices (including infection control, physical examinations, and critical incident stress management). d. training-related health and safety issues. e. development and maintenance of consistent, updated standard operating guidelines (SOGs) for operational guidance.			
4.	Define considerations for assessing the role of an occupational safety and health committee within a department's overall health and safety plan.			
5.	Present a summary of findings.			

Evaluator: _____   _____
                                    Printed   **Signed**

Comments:

_____

_____

Candidate Name: _____   Date: _____

TASK SHEET: 7.4.4 NFPA 1021, 2020 **General Requirements**	**Task:** Develop a plan for a capital improvement project or program, given an unmet need in the community, so that there is adequate information to educate citizens about the needs of the department.
PERFORMANCE OUTCOME:	The candidate will demonstrate the ability to conduct a needs assessment plan, to effectively communicate in writing, and to problem solve.

EQUIPMENT REQUIRED: Computer, pen, paper, department policies and procedures.

CONDITIONS: Given strategic planning, capital improvement planning and budgeting, and facility planning, the candidate shall:

No.	Task Steps	Date	Evaluator Score	Evaluator Initials
1.	Define the demographics for the community, its population, and its potential hazards.			
2.	Identify any customer and/or community special needs and/or unmet needs.			
3.	Define the specific capital improvement project or program and assign objectives and budget proposal.			
4.	Design an assessment tool and analysis for the present program or project.			
5.	Define how the jurisdiction will fund the capital project and program.			
6.	Present a summary of findings.			

Evaluator: _____   _____
                              Printed   **Signed**

Comments:

_____

_____

Candidate Name: _____   Date: _____

TASK SHEET: 7.4.5 NFPA 1021, 2020 **General Requirements**	**Task:** Develop a succession plan, given department resources, policies, and procedures, so that the future needs of the department are met.
PERFORMANCE OUTCOME:	The candidate will demonstrate the ability to identify employees with the abilities and desire for potential promotion, to conduct a personnel needs assessment, to effectively communicate the department's requirements for advancement, and to plan future hiring processes to meet the needs of the department to fill future vacancies.

EQUIPMENT REQUIRED: Computer, pen, paper, department policies and procedures.

CONDITIONS: Given strategic planning, member demographics, recruitment, and retention, the candidate shall:

No.	Task Steps	Date	Evaluator Score	Evaluator Initials
1.	Develop and implement an annual employee evaluation system.			
2.	Establish clear job descriptions identifying needed educational and experience objectives for each position.			
3.	Develop training program for supervisors dealing with the identification of employees with potential for future promotion.			
4.	Establish department guidelines for a fair and open mentoring program.			
5.	Develop guidelines for the training of subordinate staff through the delegation of supervisory tasks.			
6.	Establish and educate all employees on the requirements of the department's promotional evaluation system.			

Evaluator: _____   _____
                                                   Printed   **Signed**

Comments:

_____

_____

Candidate Name: _____ Date: _____

TASK SHEET: 7.6.1 NFPA 1021, 2020 General Requirements	**Task:** Develop a comprehensive disaster plan that integrates other agencies' resources, given risk, vulnerability, and capability data, so that the organization can mitigate the impact to the community.
PERFORMANCE OUTCOME:	The candidate will demonstrate the ability to analyze data, to communicate, to develop a disaster plan, and to coordinate interagency activity.

EQUIPMENT REQUIRED: Computer, pen, paper, department policies and procedures.

CONDITIONS: Given major incident policies and procedures; physical and geographic characteristics; demographics; target hazards; incident management systems; communications systems; intelligence data; contractual and mutual-aid agreements; and local, state/provincial, and federal regulations and resources, the candidate shall:

No.	Task Steps	Date	Evaluator Score	Evaluator Initials
1.	Identify the mission of department.			
2.	Identify the use of Incident Management System (IMS).			
3.	Identify the role of specialized decision-makers.			
4.	Identify delegation of authority.			
5.	Identify specific tasks of management personnel.			
6.	Identify the method of hazard assessment.			
7.	Identify the fire department operation plan.			
8.	Identify interagency cooperation and list specific agencies for response to AHJ.			
9.	Describe proposed action plan for comprehensive disaster response.			
10.	Describe special considerations for civil disturbance incidents.			
11.	Develop an executive summary.			
12.	Present a summary of findings.			

Evaluator: _____ _____
Printed   Signed

Comments:

_____

_____

Candidate Name: _____ Date: _____

TASK SHEET: 7.6.2 NFPA 1021, 2020 **General Requirements**	**Task:** Develop a comprehensive plan, given data (including agency data), so that the agency operates at a hostile event, integrates with other agencies' actions, and provides for the safety and protection of members.

PERFORMANCE OUTCOME:	The candidate will demonstrate the ability to communicate and organize a plan, familiarity with interagency planning, and coordination.

EQUIPMENT REQUIRED: Computer, pen, paper, department policies and procedures.

CONDITIONS: Given major incident plans; policies and procedures; physical and geographic characteristics; demographics; incident management systems; communications systems; contractual and mutual-aid agreements; local, state/provincial, and federal regulations and resources; and NFPA 3000 Standard for an Active Shooter/Hostile Event Response (ASHER) Program, the candidate shall:

No.	Task Steps	Date	Evaluator Score	Evaluator Initials
1.	Review major incident policies and procedures.			
2.	Assess physical and geographic characteristics.			
3.	Identify demographics.			
4.	Review IMSs.			
5.	Assess communications systems and their interoperability.			
6.	Review contractual and mutual-aid agreements.			
7.	Apply local, state/provincial, and federal laws, statutes, regulations, and ordinances.			
8.	Assess local, state/provincial, and federal resources.			
9.	Perform interorganizational planning and coordination.			
10.	Develop a disaster plan for civil disturbance.			
11.	Develop an executive summary.			
12.	Present a summary of findings.			

Evaluator: _____   _____
Printed   **Signed**

Comments:

_____

_____

Candidate Name: _____   Date: _____

TASK SHEET: 7.7.1 NFPA 1021, 2020 General Requirements	**Task:** Maintain, develop, and provide leadership for a risk management program, given specific data, so that injuries and property damage incidents are reduced.

PERFORMANCE OUTCOME:	The candidate will demonstrate the ability to communicate, analyze data, and use evaluative methods.

EQUIPMENT REQUIRED: Computer, pen, paper, department policies and procedures.

CONDITIONS: Given risk management concepts; occupational requirements; occupational hazards analysis; and disability procedures, regulations, and laws, the candidate shall:

No.	Task Steps	Date	Evaluator Score	Evaluator Initials
1.	Identify the mission of the department.			
2.	Identify risk and trends related to department health and safety issues.			
3.	Select and define risk management goals and objectives to alleviate potential concerns.			
4.	Reference appropriate laws, standards, and regulations.			
5.	Identify requirements for meeting goals and objectives.			
6.	Describe the implementations process and timeframe.			
7.	Identify a schedule for periodic evaluation of progress.			
8.	Develop an executive summary/statement.			
9.	Present a summary of findings.			

Evaluator: _____   _____
                            Printed   **Signed**

Comments:

_____

_____